Construction of Angle trisection

(An Addition in Euclidean geometry)

Manoranjan ghoshal

Copyright © 2025 by Manoranjan Ghoshal
All rights reserved.

No part of this publication may be reproduced, distributed, or transmitted in any form or by any means, including photocopying, recording, or other electronic or mechanical methods, without the prior written permission of the publisher, except in the case of brief quotations embodied in critical reviews and certain other noncommercial uses permitted by copyright law.

For permission requests, write to the publisher at:
Email: ghoshal.manoranjan@gmail.com

This book is a work of research and intellectual effort. While the author has made every effort to ensure the accuracy and completeness of the information contained herein, the publisher and author assume no responsibility for errors or omissions.

First Edition: 2025

Contents

Description..4

About Author 6

Acknowledgment..7

Introduction..8

About the problem: 10

Proof of impossibility:...12

Tools and limitation of classical geometry:..................14

Base of possibility and impossibility of construction:...16

Euclidean method of equal segmentation of field:.......18

New method of segments of field: 20

What is extended method? 24

Construction and proof:..26

Method for trisecting of any angle: 29

Another proof: 33

Description

Unlock the Timeless Mystery of Angle Trisection

For centuries, mathematicians have wrestled with one of Euclid's most tantalizing challenges — the trisection of an arbitrary angle using only a compass and straightedge. While traditionally deemed impossible within classical Euclidean constraints, this book offers a refreshing perspective and bold analytical journey into the heart of geometric construction.

In *Construction of Angle Trisection: An Addition of Euclidean Geometry*, author **Manoranjan Ghoshal** brings forth an innovative approach to this ancient enigma. With clarity, rigor, and mathematical elegance, the book introduces new constructs and logical methods that stretch the traditional boundaries of geometry — while respecting its core foundations.

Whether you're a student, teacher, or lifelong lover of mathematics, this book invites you to:

- Explore the historical context and significance of angle trisection.
- Understand the limitations of classical construction methods.
- Follow step-by-step explanations and original diagrams.

- Discover a novel solution proposed as an addition — not contradiction — to Euclidean principles.

This is more than a geometry book — it is a testament to human curiosity, mathematical beauty, and the pursuit of intellectual possibility.

About Author

Manoranjan Ghoshal, Author Born: 2nd January 1972, at Atmarampur, Kolkata, W B, INDIA.

Father's name Kartick ghoshal and mothers name Radharani ghoshal.

A writer of pure science, and literature, continue writing from many years.

Awarded from IIK, as the honorable fellow in mathematics.

Acknowledgment

First and foremost, I express my heartfelt gratitude to the timeless legacy of **Euclid** and the countless mathematicians whose work laid the foundation for geometric inquiry. Their dedication to logic, structure, and beauty continues to inspire minds across generations.

I am deeply thankful to my mentors, teachers, and fellow explorers of mathematics who nurtured my curiosity and challenged me to think beyond the obvious.

Special thanks to my family and friends, whose encouragement and support gave me the strength to pursue and complete this work. Your belief in me means more than words can express.

Lastly, to the readers — thank you for your interest, your questions, and your passion for knowledge. May this book ignite new perspectives and encourage your own journey of discovery.

– **Manoranjan Ghoshal**

Introduction

Structural Mathematics means geometry, and classical geometry means Euclidean geometry.

Our previous Mathematicians are almost decided that, classical geometry is limited and almost work had been done by Euclid self.

So, unsolved problems remaining unsolved, and they found the cause of impossibility.

But, nothing impossible! Our knowledge had been blocked for solution of those.

Now here in this book I am showing the construction for solution of Angle trisection classically.

I appeal to all Mathematicians or Geometers for see my construction carefully, after then speak out solve or not?

And throughout the challenge to all Mathematicians or Geometers, who have believed it is unsolved now.

Now, I claiming that, the construction of Angle trisection solved classically.

-----START-----

About the problem:

One of the famous Greek construction problems in classical geometry has been Angle trisection.

Although, some angles being trisection classically, such as 360°, 180°, 90°, 45° etc. but trisection of every angles was not possible classically with a general process.

Our mathematics has to need the solution.

Perhaps our previous great Mathematicians or Geometers tried to solve the problem.

They were unsuccessful.

Then they established, the problem unsolved because, it is cubic equation problem.

I strongly appeal that, are they proved exactly? Is this exactly impossible?

Proof of impossibility:

I am not interested and satisfied on their proof of impossibility. But their discovery and development of mathematics are important for us.

They developed the study of geometry is related with algebraic equation.

So, the possibility or impossibility of any geometric construction may be determined of analysis on algebraic equation.

I don't negotiate about algebraic analytic geometry.

But, the analysis having to need classically, that means Euclidean method.

Tools and limitation of classical geometry:

Construction of classical geometry need two tools, one is ruler and

Another is compass.

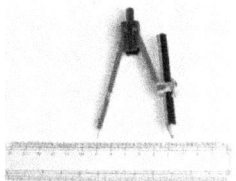

(Modern compass and ruler)

Those tools have the limitation of function.

Ruler can join two points and can extend the line as it is.

So, compass can cut segments equally.

Beyond of these function of tools never permit in classical geometry.

Great mathematician Archimedes had given a construction of angle trisection, with use of marked ruler, so that was beyond of classical geometry. My construction so closed of Archimedes, in classical method.

Base of possibility and impossibility of construction:

Greek frameworks shows in particular field the operation of equal segmentation only possible rationally.

The particular field may be rational or irrational. But the equal segmentation in it must be possible rational. (Courant and Robbins said in the book, what is mathematics?) Here is my strong objection.

Application of Euclidean equal segmentation method can create rational number of segments in field.

Now, I am showing here a method that can create rational and irrational segments in field.

Euclidean method of equal segmentation of field:

Here AB be a field, I can divide it to equal division rationally. 2,3,4,5 etc. so, could be created ½, 1/3, ¼, 1/5, etc. these all numbers are rational.

Therefore, in this method I can create only rational numbers.

A─────────────B

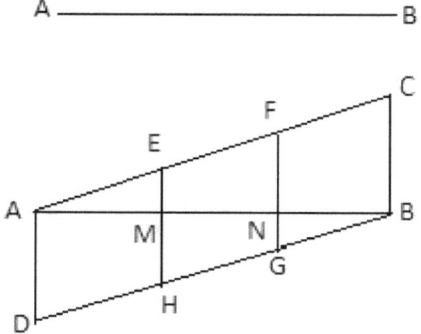

Here AB the line and AC, BD parallel, if it is AE = EF = FC, and BG = GH = HD, then will be AM = MN = NB.

New method of segments of field:

If I can apply the method of Pythagoras theorem to divide the field, then I can create the numbers in field rational and irrational, and can be continuing up to close of field.

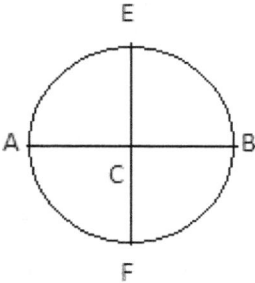

Here AB be a field, C is the midpoint of AB, draw a straight line EF through C, and cuts it AC = EC, BC = FC, now when AB and EF perpendicular at C, then AB divided to two equal parts. But while angle ACE not equal to 90°, it creates rational and irrational numbers in field.

Suppose, angle ACE = 60°, draw a perpendicular on AC from E, DE is the perpendicular, now DC = ½ of AC, suppose, AC = 1, and AB = 2, when angle ACE = 45°, then DC = 1/√2, and when angle ACE = 60°, then DC will be √3/2, now in total field

AB, others hand on BC creates same values of segments.

Here FG is the perpendicular on BC.

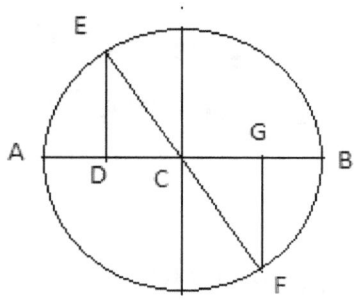

So, DG provides 1 for angle 60°, ✓ 2 for 45°, ✓ 3 for 60°, and when angle ACE = 0°, then DG will be 2 , and field will be closed. The C may be 0 and AC = -1, BC = +1. So, it creates field values from 0 to 2.

It is proving that, the ruler and compass creates not only rational fields, it creates irrational fields by rational operation.

I can construct trisection of any angle by its extended method.

What is extended method?

Suppose a diameter of circle is a field that is two of unit. And added other half circle with it, that is extended field, and the value of total field will be 3.

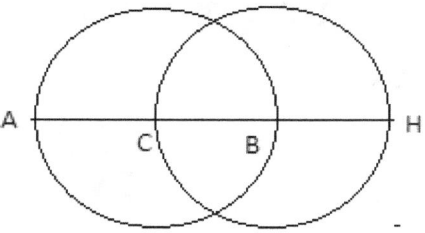

Here AB is a field and BH being extended field.

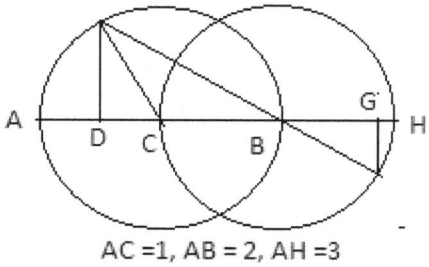

AC =1, AB = 2, AH =3

Same procedure being applying here, in two circles, so, the field value creates in it from 0 to 3. (While AC+CB+BH)

Here is the solution of Angle trisection.

Construction and proof:

AB the line, C is any point in it, drawing CD = DE and DE = EF, it create angle AEF = 3ACF angle.

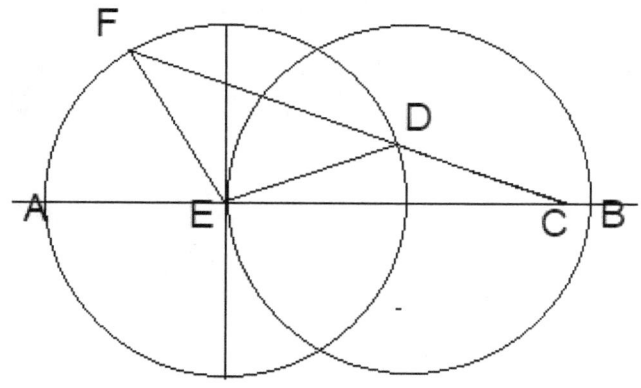

Drawing a circle center E and radius CD, it cuts to AB at G, again drawing another same radius circle from center G, joining points G,F and extending it up to H, the GH = CD, now joining two points C,H.

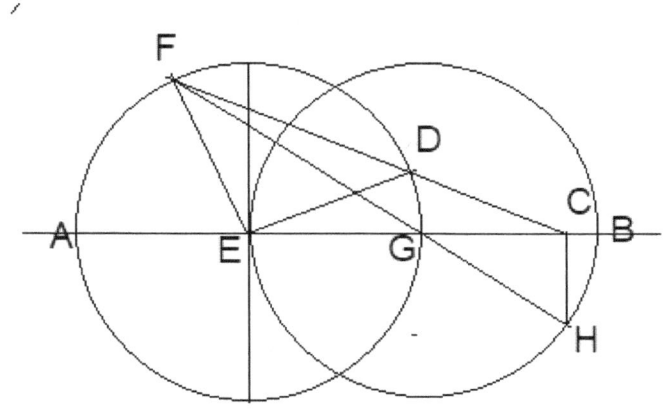

Here the CH perpendicular on AB. Suppose, angle ACH greater than 90°, when angle ACF terns to 0, then C will go to I (at the point situated in circle by the cross of AB line), but CH will not go, that means

CI and CH are not same, it is untrue.

Oppositely, if it is considering angle ACH less than 90°, then again creates same problem, only satisfied and true the angle ACH = 90°. That means CH is perpendicular on AB. So you draw FGH the straight line, and draw HC perpendicular on AB, after join C,F. here angle ECF is trisect of angle AEF.

Here is hidden the solution of angle trisection.

Method for trisecting of any angle:

Suppose, ABC is the given angle, I have to need it trisection.

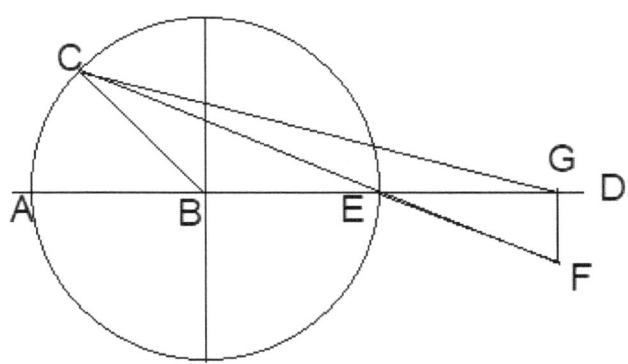

The AB extending up to E and drawing a circle of center B, joining C,E and extending it up to F, the EF = EB, the perpendicular FG is drawing on ED, now the G,C is joining.

Here, angle AGC is forming trisection of angle ABC.

Drawing, here trisections of different type angles.

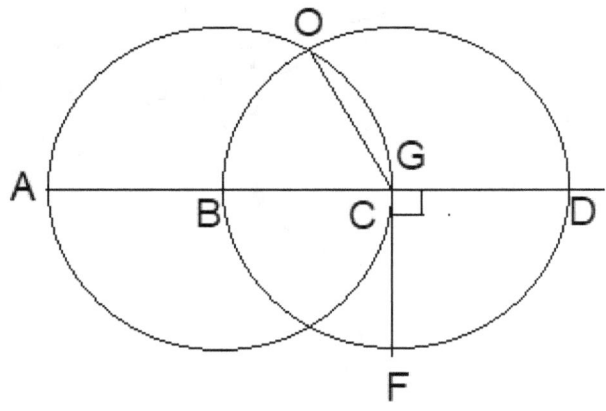

Here ABC the given angle = 180^0 a perpendicular from C

on AD is drawing CF, and GO is the radius length or GO = AB, this angle ACO = 60^0, that is $1/3^{rd}$ of angle ABC.

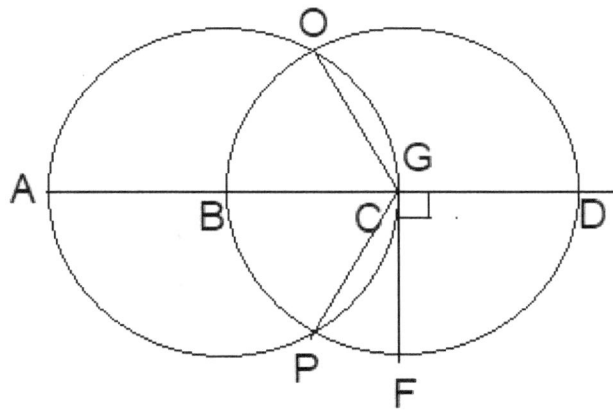

When the given angle is B = 360^0 then the angle OGP will be trisection of that angle 120^0.

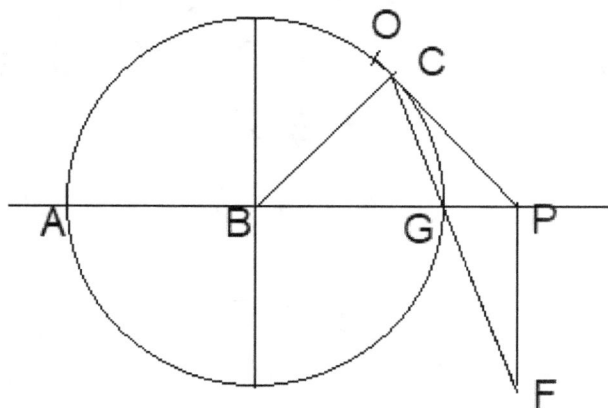

Here angle ABC is given, joining C, G up to F, and PF perpendicular on AB, the PO = AB lying on circles O.

The angle APO is trisection of angle ABC.

It is clear in this process the CG line drawing will not be possible when the given angle

ABC greater than 135^0, then you may use this process.

Another proof:

Here, AB is a straight line,

A ——————————————— B

Now, O is on a point, Drawing a circle COD,

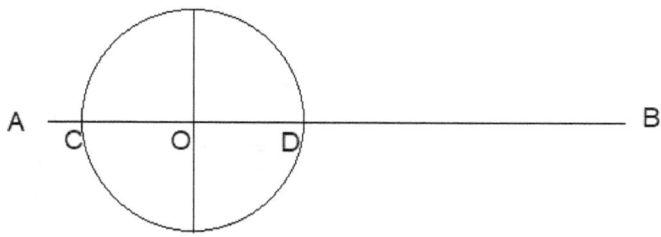

Now, put OD = EF,

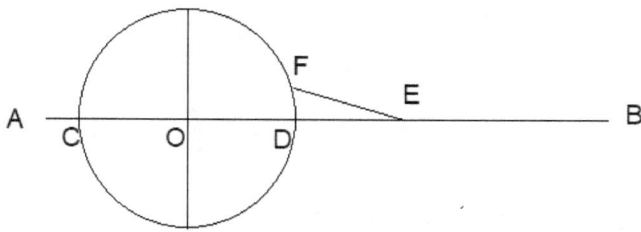

Extending it up to G, so the angle AEG is formed 1/3rd of angle AOG.

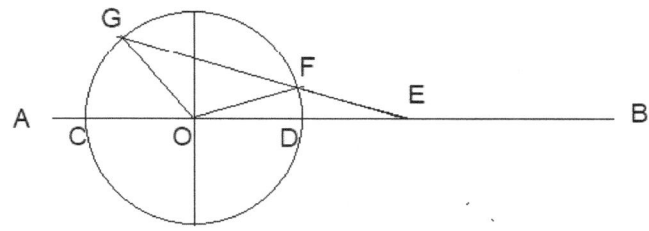

Now, drawing another circle of radius OD, O and D are two separate centers. Joining G, D and extending it up to H.

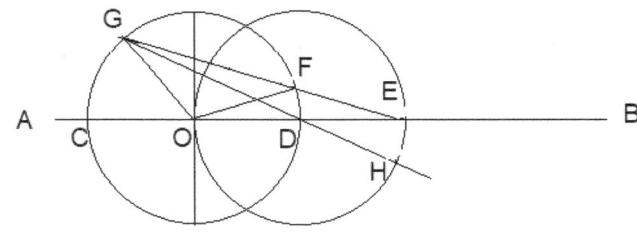

Now, joining H, E, it will be proved the HE is perpendicular on AB.

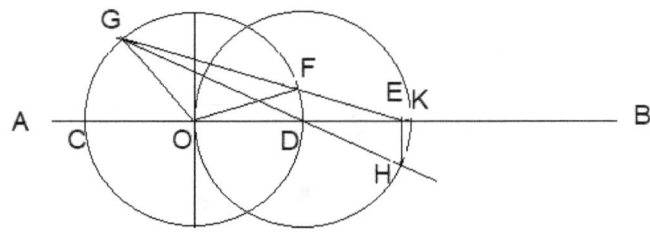

Three conditions are arrives here,

1. The angle DEH greater than 90^0.
2. The angle DEH less than 90^0.
3. The angle exact 90^0.

Proof:

It is clear here DH = FE = DK, when will go at K, then E will go at K, we know that, while the H is going to K through arc, only the perpendicular from H on AB, will go to meet at K, that means HE the exact perpendicular on AB straight line. So the perpendicular from H is the point of an end of line of trisection of angle AOG, when it will be given.

Now it is clear classical problem of angle trisection with ruler and compass is solved.

NB: here all drawing are made by assumption with eyes, it is not mathematical drawing, so you may examine the drawings by draw self mathematically.